THE CHILD'S BOOK OF THE SEASONS

LECTOR HOUSE PUBLIC DOMAIN WORKS

THE CHILD'S BOOK OF THE SEASONS

ARTHUR RANSOME

ISBN: 978-93-5614-215-2

Published: 1906

LECTOR HOUSE LLP
E-MAIL: lectorpublishing@gmail.com

THE CHILD'S BOOK OF THE SEASONS

BY

ARTHUR RANSOME

Author of "The Stone Lady."

NATURE BOOKS FOR CHILDREN.

WITH ILLUSTRATIONS BY
FRANCES CRAINE

1906

FOR ROBERT AND PHYLLIS.

CONTENTS

I

SPRING

Spring always seems to begin on the morning that the Imp, in a bright pink nightgown, comes rushing into my room without knocking, and throws himself on my bed, with a sprig of almond blossom in his hand. You see, the almond blossom grows just outside the Imp's window, and the Imp watches it very carefully. We are none of us allowed to see it until it is ready, and then, as soon as there is a sprig really out, he picks it and flies all round the house showing it to everybody. For the Imp loves the Spring, and we all know that those beautiful pink blossoms mean that Spring is very near.

"Spring!" shouts the Imp, waving his almond blossoms, and we begin to keep a little note-book, and write down in it after the almond blossom day all the other days of the really important things, the day when we first see the brimstone butterfly, big and pale and golden yellow, flitting along the hedgerow near the ground, and the day of picking the first primrose and the finding of the first bird's nest.

And then walks begin to be real fun. No dull jig-jog, jig-jog, just so many miles before going home to lunch, when all the time you would much rather have stayed at home altogether. The Imp and the Elf love Spring walks, and are always running ahead trying to see things. There are such a lot of things to see, and every one of them means that Summer is a little nearer. And that is a jolly piece of news, is it not?

The Imp and the Elf have a nurse to take them for walks, and a very nice old nurse she is, with lots of fairy tales. But somehow she is not much interested in flowers, or birds, or mice, or even in the Spring, so that very soon after the day of the apple blossom those two children start coming to my door soon after breakfast. They knock both at once very quietly. I pretend not to hear. They knock again, and still I do not answer. Then they thunder on the door. Do you know how to thunder on a door? You do it by doubling up your fist and hitting hard with the podgy part that comes at the end where your thumb is not. You can make a tremendous noise that way, And then suddenly I jump up and roar out, "Who's there?" as if I were a terrible giant. And the Imp and the Elf come tumbling in, and stand in front of me, and bow and say, "Oh, Mr. Ogre, we hope you are not very really truly busy, because we want you to come for a walk."

And then we stick our things on, and away we go through the garden and into the fields, with our three pairs of eyes as wide open as they will go, so as not to miss anything.

We watch the lark rise high off the ploughed lands and sing up into the sky. He is a little speckled brown bird with a very conceited head, if only you can get near enough to see him. The Imp says he ought not to be so proud just because he has a fine voice. And certainly, if you watch the way he swings into the air, with little leaps of flying, higher and higher and higher, you cannot help thinking that perhaps he does think a little too much of himself. He likes to climb higher than all the other birds, just as if he were a little choir boy perched up in the organ loft. He climbs up and up the sky till you can scarcely see him, but he takes care that you do not forget him even if he is so high as to be out of sight. He sings and sings and sings. The Imp and the Elf like to wait and watch him till he drops down again in long jumps, just as if he were that little choir boy coming down the stairs ten steps at a time. "Now he's coming," says the Imp, as he sees the lark poise for an instant. "Now he's coming," the Elf cries, as he drops a foot or two. But we always think he is coming before he really is.

As we go through the fields we keep a good look out for primroses and cowslips. The primroses come long before the cowslips. Cowslips really belong to the beginning of the Summer. But early in the Spring there are plenty of woods and banks we know, where we are sure of finding primroses with their narrow, furry, green leaves, and the pale yellow flowers on a long stalk sprouting out of the heart of the leaves.

In the primrose-wood where we always go in the Spring, we find lots and lots of primroses, and some of them are not yellow at all, but pale pink and purple coloured. The Elf collects them for her garden, and she carries a little trowel and digs deep down into the earth all round them so as not to hurt their roots and then pulls them up, and puts them in the basket to plant in her garden at home. You see, they really belong to gardens, for they are not quite proper primroses, but the children of primroses and polyanthuses. You know polyanthuses. The Imp says their names are much too long for them. But you know them quite well, just like cowslips, they are, only all sorts of colours.

About the same time that the primroses are out the wild dog violets begin to show themselves. We always know when to look for them, for wild ones bloom as soon as the sweet ones in our garden are over. The Elf watches the garden violets and picks the last bunch of them, and ties them up with black cotton and puts them on my plate ready for me when I come down late to breakfast. Yes, I do come down late for breakfast. I know it is naughty, but you see even grown-ups are naughty sometimes. The Imp thinks I am very naughty indeed, and so one day, when I was late, he took my porridge, and got on a high chair, and put it on the top of the grandfather clock for a punishment. You see, whenever the Imp and the Elf are late they have to go without porridge. That is why they are very seldom late. Well, as soon as I came down I saw my blue porridge bowl smiling over the top of the clock, and I just reached up and took it down and ate it, and very good porridge it was, too. But the Imp said, "It's horrid of you, Ogre, to be so big," and then he laughed, and I laughed, and it was all right.

Oh, yes, I was just telling you that the Elf put the last bunch of the sweet violets on my plate. Well, when that happens we all know that our next walk will be to the places where the wild violets grow for they are sure to be just coming out.

The wild violets are just like the sweet ones in liking cool, shady places for their homes. We find them nestling in the banks under the hedge that runs along the side of the wood. They cuddle close down to the ground, with their tiny heart-shaped leaves and wee pale purple flowers, just like little untidy twisted pansies.

The Elf reminds me that I am to tell you about the daffodils. I had forgotten all about them. Really, you know it is the Imp and the Elf who are writing this book. If it were not for them I should be forgetting nearly everything. There are such a lot of things to remember In the wood where we find the coloured prim-roses there are great banks of daffodils under the green larches. They are just like bright yellow trumpets growing out of pale yellow stars. The Imp says they are the golden horns the fairies blow when they go riding through the woodland in the moonlight on their fairy coaches. I do not know if he is right, but anyhow they are very pretty. They have lots of long flat leaves growing close round each flower, like sword blades sticking up out of the ground, and the buds look at first as if they were two leaves tightly rolled together. And then the green opens and a pale spike comes out, and a thin covering bursts off the spike, and the spike opens into the five-pointed star, leaving the brilliant golden trumpet in the middle. Gardeners, and that sort of person grow double daffodils that look like two flowers one inside the other, but the ordinary wild daffodil is far the prettier. At least the Imp and the Elf think so, and I think so, too.

We go to the wood and lie down on the dried leaves from last year, and watch the flowers and talk about them and the little mice who live in the undergrowth. Sometimes, if we are not too lazy, the Elf makes us pick primroses and daffodils and violets to send to children we know in town—pale-faced children who think we must be dull in the country, with nothing to do, and no pantomimes. Really, of course, there is such a lot to do in the country that we have always got the next thing planned before we have done what we are doing. And as for pantomimes this very wood is just like a theatre, with mice and rabbits and birds for actors, and the most beautiful transformation scenes. Why, just now in Spring it is yellow with primroses and daffodils, with pale larches wearing their new green dresses. But soon all the trees will be green, and the whole wood will be carpeted in blue, deep rich blue, the colour of the wild blue-bells, whose leaves we can see coming up all over the place. Spiky green leaves they are, and the children see them at once. "Blue-bells are coming," sing the Imp and the Elf, and so they are, and with the blue-bells comes Summer.

Besides the lark and cuckoo, who is going to be talked about in a minute, be-sides the flowers, there are other things we watch for signs of Summer, and those are the trees themselves.

We watch the trees for flowers and for buds. From the high windows of the house we can see over the fields to the woods, and see them change colour from the dead bareness of Winter very early in the Spring. And when we go to the woods in daffodil time we all three of us watch the buds coming out on every branch farthest out on the lowest boughs, which for Imps and Elves are also easiest to see.

Earlier than this we look for catkins on the hazel trees. The Elf calls the hazels "the little children of the wood" because they grow low, and the other trees, the oaks, and beeches, and elms, and chest-nuts, and birches, tower above them. In some parts of the country catkins are called lambstails, because they hang down just like the flabby little tails of the Spring lambs. What do you think they really are? The Elf would not believe me when I told her they were hazel flowers. "Trees don't have flowers," she said. I reminded her of hawthorns and wild roses, and she said, "Oh, yes, but these things are greeny-brown and not like flowers at all." But they *are* flowers. They are the flowers of the hazel tree, and they are almost the very first of the Spring things that we see. If you look about when you are in the woods you will find that lots of other trees have green flowers, too, and many of them just the same shape as the lambstails.

The Imp and the Elf are early on the look-out for another tree-flower that is one of the Spring signs, and that is the flower that people who know nothing about it call "palm." Hundreds of men and women from the towns come out into the country to gather it, and a horrible mess they make of our country lanes and fields. The Elf calls them the "Ginger-beer-bottle-and-paper-bag-people" and hates them with all her small heart.

Really, that flower that those people come to gather belongs to the sallow, which is a kind of willow. You know it quite well, with its beautiful straight, tall, bendable stems that look as if they were simply made for bows and arrows. In Spring-time the sallow flowers in pretty little silvery tufts, soft and silky to touch, clinging all along its twigs. The Elf always picks the first bit that she can find that is really out and carries it home in triumph, and puts it in a jampot full of water to remind her that Winter is really over and gone.

On the way to the woods we have to pass through broad green fields full of grey sheep with long tangled wool all nibbling at the grass. And very early in the Spring a day comes when by the side of one of the old grey sheep there is some-thing small and white. And the Elf says nothing, but slips her hand into mine, so that I can feel it shaking with excitement. She touches the Imp, so that he sees the white thing, too, and then we all three go across the field as quietly as ever we can to see the little new lamb as near as possible. But little lambs and their grey moth-ers are very nervous, and long before we are really close to them the grey sheep moves away, and the little white lamb jumps up and scampers after her.

Before the Spring is half through nearly all the grey sheep have one or two little white woolly children trotting about with them, and we watch the lambs chasing each other and skipping over tussocks of grass like little wild mountain goats. The Imp and the Elf are always wondering what they think about in those queer little heads of theirs, with the big ears and great round puzzled eyes.

But of all the Spring signs the oldest and sweetest and dearest is the cry of the cuckoo that comes when Spring is just going to change into Summer. For hun-dreds of years English children have listened for the cuckoo in the Spring, and the very oldest English song that was ever written down is all about the cuckoo's cry.

"Summer is a coming in,
Loudly sing cuckoo.
Groweth seed and bloweth mead,
And springeth the wood now.
Ewe bleateth after lamb,
Lowth after calf the cow,
Bullock starteth,
Buck now verteth,
Merry sing cuckoo.
Sing cuckoo,
Well singeth thou, cuckoo, cuckoo.
Nor cease thou ever now."

The Imp and the Elf love that little song and know it by heart. It was written by an old monk in the Spring-time years and years and years ago, and some of the words he used are difficult to understand now. Verteth is an old word meaning going on the green grass. Nearly all the other words I have made as much like our own as I can.

It is much easier to hear the cuckoo than to see him. He is a biggish grey gentleman with stripes across his middle, and he is horribly hard to notice unless we get quite close to him. He is very shy, and that makes it harder still. But sometimes when you hear him cry, cuckoo, cuckoo, if you are very quick indeed, you can see him flying across a field from hedge to hedge.

Mrs. Cuckoo is the laziest mother that ever was. The Elf thinks her perfectly horrid. I wonder if you know why? She is so gay and fond of flying about that she finds she has no time to build a nest or bring up her little ones as all good mothers do. So she just leaves her egg in someone else's home, and flies happily away, leaving the someone else to hatch the egg and bring up the little cuckoo. She often chooses quite small birds like the little greenfinches or even the sparrows. And when the young cuckoo comes tumbling out of his egg, instead of being kind and polite to the children to whom his nursery really belongs he just wriggles his big naked body under them and tumbles them out of the nest. That is why, though we love to hear the cuckoo, we think him rather a lazy bird, and his wife a very second-rate kind of mother.

When we come back from the walk on which we have heard the first cuckoo of the year, we really begin to long for the Summer. All the Spring signs have come. When we get back to my room, the Imp and the Elf sit on my table and swing their legs and say, "Brimstone butterfly, palm, catkins, daffodils, violets, primroses, blue-bells, and cuckoo; Summer is coming, don't you think, Ogre?" And I say yes. And they say, "Tell us what Summer is like, do, please." And I tell them, though they know already, and they sit on the table and wriggle at all the nicest parts of the telling, and we are all very happy indeed.

II

SUMMER

And what are the things we know the Summer by? Summer clothes say little girls, and big straw hats say boys. Well, and what do they mean but the heat? The Imp wears a huge straw hat and a loose holland overall but he goes about panting, and lies flat on the ground with the straw hat over his nose to keep the sun from burning his face. And the Elf wears an overall, too, and a pale blue calico sunbonnet over her curls. All the same she is often too hot to enjoy anything except sitting in the swing in the orchard and listening to fairy tales. And I, well, I am often too hot to tell fairy tales. For fairyland is a cool, comfortable place, and big, hot Ogres melt it away like an ice palace.

"Yes, yes, yes," you say, "but what do you do? It can't always be too hot to do anything." I asked the Elf what we do do in Summer time, and her eyes grew bigger and bigger, and she clapped her hands and said, "Do? Why, everything." And now I am going to try to tell you a few of the things that make the everything so delightful.

First of all there are cowslip balls. We go, the three of us, to the field where the cowslips grow. Little cousins of the primroses the cowslips are, as you know already. Well, we take a long piece of string and fasten one end to a bush, and pick piles of flowers close to the top of the long stalk and hang them over the thread, so that some of the flowers hang on one side and some on the other. And when we have a great row hanging on the thread we take its two ends and tie them together. And all the cowslips tumble into the middle and crowd up against each other, and when the thread is tied they are packed so close that they make a beautiful ball, with nothing but cowslip faces to be seen all over it. And that is a cowslip ball.

Close under the moor, not so very far away from the house, there is a gate where the lane divides into three or four rough paths that run over the heather to the moorland farms. And just by the gate there is a hawthorn tree. The flowers of the hawthorn are not, like the catkins, over before the hazel shows its leaves. They wait till all the tree is vivid green, and then sparkle out all over it in brilliant white or coral colour. We call the hawthorn May. And a long time ago all over England on May-Day people used to pick the May and make a crown of it and decorate a high pole in the middle of each village. And then they danced round the pole, and crowned the prettiest of the girls and called her the Queen of the May. She had a sprig of hawthorn blossom for a sceptre, and everybody did what she told them. It must have been rather nice for the little girl who was chosen Queen.

But now nearly everybody has forgotten about May-Day fun. Perhaps they would not enjoy it even if they remembered. But here, when the May is out, the country children from the farms over the moorland and from this end of the valley choose a fine day and come to the tree. The Imp and the Elf always take care to find out when they are coming. Then they bang on the study door for me and away we go, with plenty of buns and sandwiches in our pockets. And always when we get to the tree we find that some of the country children are there before us. And soon the fun begins. They all dance round the tree, and after eating all the buns and things they choose a King and Queen, and play Oranges and Lemons, the King and Queen leading off. This year they chose the Imp and the Elf, and you just can't imagine how proud they were, and how the Imp strutted about with his hawthorn sceptre, and the Elf kept re-arranging her curls under her green and starry crown. The sun shone all day, and we were all as happy as anyone could wish to be.

Then, too, in Summer we go quietly and softly through the little wood at the back of the house and wait at the other side of it and peep over the hedge. There is a steep bank on the other side and then a row of little trees, the remains of an old hedge, and then another bank. And the other bank is full of holes, and the holes are full of rabbits. And in the Summer evening we go there and watch the little rabbits skipping about and nibbling the grass. And of course as the Summer goes on the grass grows very high, and when we walk through it we can sometimes see nothing but the ears of the little rabbits peeping up above it. You can't imagine how funny they look. Once the Imp fell right over the top of one of them that was hidden in the grass. It jumped out under his feet and he was so startled that he fell forward, and felt something warm and furry wriggling in his hands, and found that he had caught a baby rabbit. The Elf and the Imp patted and stroked it till it was not frightened any more, and then we put it on the ground and let it go. It hopped gaily away through the grass and disappeared into its burrow in the bank. I do not wonder that it was a little afraid and trembly when the Imp, who must seem a giant to it though he only seems a boy to me, came bumping down on it out of the sky.

Besides the rabbits we find all sorts of other charming things in the long grass that swishes so happily round our ankles. Buttercups are there which send a golden light over your chin if you hold them near enough, buttercups, and dandelions, and purple thistles, and wild orchids. You know thistles and dandelions, of course, but I wonder if you know an orchid when you see one? They are quite common things, but lots of even country children do not bother to look for them. Next time you are in the fields in Summer just look about you for a spike of tiny purple flowers with speckled lips rising out of a little cluster of green leaves with brown spots on them. Soon after these have begun to flower we often find another kind, with speckled flowers too, only far paler purple. And later still there is a meadow where we can usually discover just a very few Butterfly orchids. They have a spike of delicate fluttery flowers, not so close together as the purple kinds, and with green in the veins of their white petals. They are a great prize and the Elf always picks one, leaving the rest, and brings it very carefully home and keeps it in water for as long as she can for it is a treasure indeed.

In another bank, not so very far from the home of the rabbits, another little furry creature lives, a pretty little brown-coated, long-tailed person, a great hunter, and much feared by the rabbits. He has a long, thin body, and a sharp little head, and a wavy tail. He is a weasel. His bank is just by the side of a pleasant little trickling beck, and not very far from the wood where the pheasants live. Some day he will be shot by the keeper for I am afraid he is rather fond of pheasant. There are plenty of stories about him among the country people. They say that if you whistle near his hole he will come running out to see what is the matter; and if you go on whistling he will come nearer and nearer until you can catch him with your hands. I have never tried, so I do not know if this is true. But I should not like to catch him in my hands for his teeth are as sharp as a rat's. At any rate there is one thing that is far more certain to bring him out of his hole than any whistle, and that is want of rabbit. Once, as we walked through the fields in the Summer twilight, we heard a short squeal and saw a poor little rabbit hopping feebly away with Mr. Weasel running nimbly along after him. And the funny thing is that the rabbit instead of scampering away as fast as he could go, was going quite slowly, and in the end stopped altogether, when the weasel ran up and killed him. The Elf said it was cruel of the weasel and silly of the rabbit. The Imp said he did not know about the weasel, but the rabbit deserved to be killed for being so slow in getting away. But our old gardener, who is wisest of us all, says that the weasel has to kill rabbits to keep alive, and that it isn't the rabbit's fault that it cannot run fast. He says that when a rabbit is chased by a weasel it cannot help going slower and slower, and being terribly frightened because it knows that it cannot escape.

The sheep in the fields are just as interesting as the rabbits or the weasels. One of the most exciting of all the Summer things has to do with them. Towards the end of May the Elf and the Imp are always bothering the farmers round about, to find out when shearing and washing time is going to be. There is an old rhyme that the farmers' wives tell us, and it says:

> "Wash in May,
> Wash wool away.
> Wash in June,
> The wool's in tune."

What it means is that if the farmers wash the sheep in May the wool is not so strong and healthy as it is later, and comes off in the water. While, if they wash them in June the wool is quite crisp and stands on end, and so is very easy to cut. Usually the farmers wash the sheep in the beginning of June, and these two children always manage to find out the day before. They come and bang at my door as usual, but when they come in they bring my hat and walking-stick with them, for they know quite well that on sheep-washing day I am sure to want to come with them. And then off we go along the road that leads by the hawthorn tree, over the moor and down on the other side, to where a little river runs between two farms about a mile from each. The river widens into a broad shallow pool, and here the farmers bring the sheep. The Imp and the Elf and I have a fine seat in the boughs of a big oak that overhangs the water. Here we sit in a row and have a splendid view. The sheep are all crowded together in pens at each side of the water, and the farmers wade out into the stream taking the sheep between their legs, and wash them one after the other. I told you there were two farms who use this washing place. Well, every year they race in their sheep-washing, and see who can get most sheep washed in the shortest time. The Elf takes one side and the Imp takes the other, and it is really quite exciting.

When the sheep are all clean they are turned loose in the fields again for a whole week in which to get properly dry. Then they are driven into the farmyards or into pens in the fields, and the farmers clip the wool off them close to the skin. They only shear the old sheep and last year's lambs; and that is why, after shearing time, the new lambs have so much finer and longer coats than those of their mothers. We always wondered why that was, until we found out that the farmers only clip the tails of the lambs, but leave their coats on, while they take all the wool off the old sheep, and send it away to be made into nightgowns and things for Imps and Elves and you and me and everybody else.

After the sheep-shearing comes the haymaking, and that is the piece of Summer fun that the Imp loves best of all. We watch the grass growing taller and taller, till the buttercups no longer tower above it, and the orchids die away. We notice all the different grasses, the beautiful feathery ones that the fairy ladies use as fans in the warm midsummer nights, and those like spears, and those like swords, flat and green and horribly sharp at the edges. We see them all grow up and up, and change their colour under the Summer sun. And then at last comes the day when the farmers of the farm across the meadow harness old Susan, the big brown horse, to a scarlet clattering rattling mowing machine that glitters in the sunlight. And then we hear it singing down the field, making a noise like somebody beating two sticks together very fast indeed. As soon as we hear that, we climb through the hedge at the bottom of the garden or over it, and run round the field, because if we came straight across it, we should trample the grass, and then the farmer would not smile at us so pleasantly. And we shout for Dick the labourer who sometimes lets the Imp or the Elf ride Susan home from the fields. We find him sitting on the little round seat at the back of the mowing machine holding the long ropes which do instead of reins. And Susan is tramping solidly ahead, and the machine drags after her, and the hay falls behind flat on the ground in great wisps. And the Imp runs along by the side of the machine, and tells Dick that he is going to be a farmer, too, when he is big enough. Do you know I never met a little boy yet, who did not want to be a farmer when the hay is being cut? I was quite certain that I was going to be a farmer myself, long ago when I was a little boy, and not an Ogre at all.

And then, when the hay is cut we toss it and dry it, and that is even jollier than the cutting. The farmer's daughters and Dick's wife come into the field and join in tossing the grass and turning it over with long wooden hayrakes, until it is quite dry. And they laugh at the Elf and the Imp, and throw great bundles of hay over the top of them. And the Imp and the Elf throw the hay back at them, and tease them until they are allowed to do a little raking and tossing for themselves. But they soon tire of that. Presently the Imp throws a wisp of hay over the Elf, and the Elf throws hay over the Imp, and then they throw hay over each other, as much and as fast as they can. And then they creep up to me, where I am sitting as a quiet comfortable Ogre should, smoking or reading. And suddenly it seems as if all the hay in all the world were being tumbled over my head and shoulders. The Imp and the Elf cover me all over with hay, and then sit on top of me, and pretend that I am a live mountain out of a fairy tale, and that they are giants, a giant and a giantess taking a rest. And suddenly the live mountain heaves itself up in the middle, and upsets the giant and the giantess one on each side. And then we all get up covered with hay and very warm, and laugh and laugh until we are too hot to laugh any longer.

When the sun falls lower in the sky, and the hedge throws a broad cool shadow over the clipped grass, we all sit under it out of the sunlight, the farmer's daughters in their bright pink blouses and blue skirts, the farmer and Dick and I smoking our pipes together, and the Elf and the Imp in their holland smocks. We all meet under the shadow of the hedge as soon as we see two figures leave the farmyard and come towards us over the fields. One of them is the maid who helps in the farm and the other is the farmer's wife. Each of them carries a big round basket. They come up to us blinking their eyes against the sun, red in the face, and smiling and jolly. And we help them to unload the baskets, which are full of food and drink. Great big slices of bunloaf, dark and full of soft, juicy raisins, and tea, the best tea you ever tasted, for tea never tastes so nice as when you drink it under a hedge and out of a ginger beer bottle.

Haymaking is better fun than sheep-washing and lasts longer. It is not all over in a day. There are such a lot of things to do. The hay has to be cut and tossed and dried and piled into haycocks before it is ready to be heaved high on pitchforks on the top of a waggon that is to carry it away to the farm. And after all when you have made hay in one field you can go and make it another. And there are such a lot of fields about here.

But when the hay is dry and ready in big, lumpy haycocks, the Imp and the Elf shout for joy to see old Susan harnessed to the big waggon come lumbering into the field, and to see the men throw the huge bundles of hay into the cart. One man stands amid the hay in the waggon and takes each new bundle as soon as it is pitchforked up, and packs it neatly with the hay already there. The hay rises higher and higher in the waggon, and the Imp loves to be in the waggon with the man, and to climb higher with the hay until at last he is high above the hedges, for by the time the cart is fully loaded you would think it was a great house of hay, ready to topple over the next minute. But the men do not seem to be afraid of that. They just fling a rope across the top and fasten it to keep all safe. And the Imp lies flat on his stomach on the very top of everything, and hears the farmer below him sing out, "Gee hoa, Susan!" and Susan swings herself forward and with one great jerk starts the waggon, and the Elf waves her pocket-handkerchief as they go rumbling away across the rough field and out on the lane to the farm, taking the hay that is to keep the cows fat and healthy through the winter.

When the last of all the carts of hay has rumbled away like that, Dick's wife, who knows all the old songs, reminds the Elf and the Imp of this one;

> "With the last load of hay
> Light-heart Summer trips away,
> When the cuckoo's double note
> Chokes within his mottled throat,
> Then we country children say,
> Light-heart Summer trips away."

Though Summer does not go quite yet, there is a sad sound already in the woods, for the cuckoo who told us that Summer was coming, sings cuckoo no longer, but only a melancholy cuck, cuck, as if he were too hot and tired to finish his song. And that means that he is going soon.

> "Cuckoo, cuckoo,
> How do you?
> In April
> I open my bill;
> In May
> I sing by night and day;
> In June
> I change my tune;
> In July
> Away I fly;
> In August
> Away I must."

It is very hard to tell when Summer ends and Autumn begins. But as soon as we hear the cuckoo drop the last note of his song we know that we must soon expect the time of golden corn, and after that of crimson leaves and orange. And when we hear the cuckoo no more at all, the Imp and Elf and I take marmalade sandwiches and bottled tea for a picnic in the hazel wood that is now thick with leaves, and so thickly peopled with caterpillars that they tumble on the Imp's big hat and get entangled in the Elf's hair as we pick our way through the trees. We have our picnic on a bank there, the very bank where we find violets and primroses in the spring, and the Elf lying close to the warm ground whispers "Good-bye, Summer," and even the cheerful Imp feels a little sad.

III

AUTUMN

I told you that we are sad when we know that Summer is passing away; but that is only because we love the Summer, with her gay flowers and fair green clothes, and not because we do not like the Autumn. The Imp and the Elf laugh at me when I tell them that all Ogres and Ogresses, all people who are grown up and can never be Imps and Elves again, love the Autumn and the Spring even better than the Summer herself. And then, to make them understand, I tell them a fairy story, how, once upon a time, Spring and Summer and Winter and Autumn were four beautiful little girls. Winter wore a white frock with red berries in her hair; Summer was dressed in deep green, with a crown of hawthorn and blue hyacinths; and Spring had a dress of vivid green, the colour of the larches in the woods, and a beautiful wreath of primroses and violets on her head; while Autumn was only allowed Summer's old dresses when they were faded and nearly worn out Autumn was very unhappy, for she loved pretty dresses like every little girl. But she went about bravely, with a smiling brown face, and never said anything about it. And then one day a fairy Godmother, just like Cinderella's, came into the garden, and asked to see all her little Godchildren. And Spring and Summer and Winter all put on their best frocks and came to be kissed by her. But poor Autumn could only tidy up Summer's old dress, which she did as well as she could, and then came out after the others. But she was shy because she knew that her dress was only an old faded one, and not so pretty as the spick and span clothes of her sisters. So she hung back and was kissed last of all. The Godmother kissed the others on the forehead, but when she came to Autumn she saw that all was not quite well, and looked at her very tenderly and said, "Tell me all about it," just as all the nicest fairy Godmothers do. And Autumn whispered that she was sorry that she was not looking as pretty as the others, but that she really could not help it, because she had no frocks of her own.

The Godmother laughed, and took her in her arms and kissed her on the lips. And then the Godmother put her arm round Autumn's neck, and, walking hand in hand, they went together down the garden under the bending trees to the edge of the pond. "Look into the water, my dear," said the Godmother, and Autumn looked and knew that a magic had been done, for her old faded dress was red and gold, and a rich gold crown lay on her dark hair. And she turned to thank the Godmother, and found that she had gone. She only heard a little laugh in the air, and then she laughed, too, and went away singing happily over the green grass. She has been happy ever since.

And really that is a true tale, and it happens once every year. You can see it happening for yourself after the end of the Summer, just as the Imp and the Elf and I watch it in the fields and woods. First you will see that Autumn is wearing Summer's old clothes, getting shabbier and shabbier and shabbier, and then the fairy Godmother comes, and you see the dusty green grow dim and dark, and then blaze in scarlet and orange, and even before this you will have seen the green corn pale and turn to deepening gold. And when these things have happened you can be very happy, for you will know that Autumn is smiling happily to herself, for she has her own dress at last.

The cutting of the golden corn is almost as jolly as the haymaking, so think the Imp and the Elf. Not quite so jolly, but very nearly. As soon as the hay is cut and tossed and dried and carted away to the stacks we begin watching the corn turn yellower and yellower while its golden grains hang heavily down. And at last, when the fields are bright gold in the sun, and the sky promises us clear weather for a day or two, the scarlet machine comes out again, and Susan has more work to do. This time it is not the hay, but the tall corn that falls swishing (with a noise just like that) behind the machine. And men go behind and bind it into corn-sheaves, great big bundles of corn, and then the corn-sheaves are piled into corn shocks. Eight sheaves stand on end in two rows of four leaning on each other. In some parts of the country they only have three in each row. As soon as the shocks are made the Imp has some delightful games. He loves to lie flat on the stubbly ground, and wriggle his way into the tunnel between the sheaves of corn until he crawls out at the other end covered with little bits of straw and prickling all over. The Elf does not like this part of it quite so much, but she does it sometimes, and once, when I was littler, I used to do it, too. But that was a very long time ago.

The girls from the farm come into the field to pick up all the stray corn that the men have dropped in making and carrying the sheaves, so that none is wasted at all. That is called gleaning. A long time ago rich farmers used to let poor women come into their fields and keep all the corn that they could glean, all that the reapers had left. In those days, instead of one man sitting on a scarlet reaping machine, they had many reapers, who took the corn in bundles in their arms and cut it off close to the ground with a curved knife called a sickle. This used to be done everywhere till the machines came, and even now there is a little farm we know over the hills where they use the sickle still.

Autumn is the gathering-in time of all the year. In Spring the farmers sow their corn. It grows all the Summer and in Autumn is harvested. In Autumn we gather the garden fruit. In Autumn we pick blackberries, and is not that the finest fun of all the year?

We go blackberrying with deep baskets, and parcels of sandwiches and cakes. We have several good blackberry walks. One of them takes us past the hawthorn tree and along the edge of the moors, and then down into a valley through a long lane with high banks covered with bram-bles and black with the squashy berries. As we pass the hawthorn tree the Elf always look up at it, and though she says nothing I know she is thinking of the Mayday and the dancing and the playing at Oranges and Lemons.

We have a basket each when we go blackberrying, and we race to see who can pick most blackberries. It is a curious thing that the Elf always wins, though the Imp and I work just as hard. Partly I think it is because little girls' fingers are so nimble. Perhaps from making doll's clothes. What do you think?

You see just grabbing blackberries is no use at all. We have to pull them care-fully from their places, so as to get the berries and nothing else; just the soft black lumps that drop with such nice little plops into the baskets, and go squish in the mouth with such a pleasant taste. Oh, yes, pleasant taste, that reminds me of an-other reason why when we get home we always find the Elf's basket more full of blackberries than the Imp's. The Imp is like me, and eats nearly as many as he picks. Blackberries are easier to carry that way.

Away behind the house there is an orchard, where there are pears and dam-sons and apples and quinces, all the very nicest English fruits. And all along the high wall of the orchard on the garden side grow plums, broad trees flat against the wall fastened up to it by little pieces of black stuff with a nail on each side of the boughs.

When the Autumn comes the Imp and the Elf slip slily round the garden path to the plum trees and pinch the beautiful purple and golden plums and the round greengages to see if they are soft. For as soon as they are soft they are ripe, and as soon as they are ripe comes picking time. And then the old gardener comes with big flat baskets and picks the plums, taking care not to bruise them. And the Imp and the Elf help as much as he allows them and he gives them plums for wages. And then they come to my study with mouths sticky and juicy with ripe plum bringing a plum or perhaps a couple of greengages for me. "For Ogres like plums even if they are busy," says the Elf, as she sits on my knee and crams half a plum and several sticky fingers into my mouth.

Then come the joyous days of apple-gathering and damson picking. When we sit on the orchard wall eating the cake that the cook sends out to us for lunch in the morning we wonder and wonder when the damsons will be ready. Long ago they have turned from green to violet, and now are deep purple. And the Imp wriggles down from the wall and climbs up the easiest of the trees and shouts out that they are quite soft. And at last the tremendous day comes when the gardener, and the gardener's boy, and the cook, and the kitchenmaid, and the housemaid all troop into the orchard with ladders and baskets. And the gardener climbs a ladder into the highest apple tree and drops the red round apples into the hands of the maids below. The Imp and the Elf seize the step-ladder from the scullery and climb up into a beautiful little apple tree that has a broad low branch that is heavy with rosy-cheeked apples. They wriggle out along the branch and eat some of the apples and drop the rest into the basket on the ground beneath them. And other people pluck the damsons from the damson trees and soon the baskets are full of crimson apples and purple damsons, and away they go into the house where the cook takes a good lot of them to make into a huge damson and apple tart that we shall eat to-morrow.

The old gardener then takes the longest of all the ladders and props it up against the quince tree, for the quince is the highest of all the trees in the orchard. One of the maids climbs half-way up below the gardener and he gathers the green and golden quinces and passes them to her and she passes them to the kitchenmaid, who stands ready on the ground to put them into the basket. And the Imp and the Elf sit in their apple tree and eat apples and laugh and then pretend that they are two wise little owls and call tuwhit tuwhoo, tuwhit tuwhoo, till I take up a walking-stick and pretend to shoot them, and then they throw apples at me and we have a game of catch with the great round red-cheeked balls. Oh, it is a jolly time, the apple-gathering, as the Imp and the Elf would tell you.

About now the fairy Godmother works her miracle for Autumn. We look up to the moors and find them no longer dark and dull for the green brackens have been turned a gorgeous orange by the early frosts in the night-times. And when we look over the farm land to the woods we find the trees no longer green, and the Elf says, "Do you see Ogre, the Godmother has crowned Autumn now?" and sure enough, for the leaves are turned like the brackens into a glory of splendid colours. And then we go and picnic in the golden woods, and sometimes when the sun is hot we could almost fancy it was summer if it were not for the colour. We picnic under the trees and do our best to get a sight of a squirrel, and watch the leaves blowing down from the trees like ruddy golden rain. Once, before we went to the woods, I asked the Imp and the Elf which leaves fall first, the leaves on the topmost branches of the trees or the leaves on the low boughs, The Imp said the top ones, the Elf said the low, and the Elf was right, although I do not think that either of them really knew. Usually the Imp is right, and when I said the low leaves fell the first, he said, "But isn't the wind stronger up above, and don't the high leaves get blown hardest?" Yes, one would think that the high leaves would be the first to drop. Can you guess why they are not? Shall I tell you? Well, you just remember in Spring how the first buds to open were the low ones. Then you will see why they are the first to fall. They are the oldest. When we came to the woods we found that this was just what was happening. All the leaves at the bottom fall, and sometimes many of the trees in the woods kept little plumes of leaves on their topmost twigs after all the others had gone fluttering down the wind.

I am only going to remind you of three more things that belong to Autumn. And one of them is pretty, and one of them is exciting, and one of them is a little sad.

The first is a garden happening. Close under the house we have a broad bed, and for some time before the real Autumn it is full of a very tall plant with lots and lots of narrow dark green leaves. And after a little it is covered with buds, rather like daisy buds only bigger. And then one day the Imp leans out of the window and gives a sudden wriggle. "Come and hold on to my legs, Ogre, but you must not look." And I hold on to his round fat legs and keep my eyes the other way. And he leans farther and farther out of the window, puffing and panting for breath, until he can reach what he wants. And then the fat legs kick backwards, and I pull him in, and when he is quite in he says, "The first," and there in his hand is a beautiful flower like a purple daisy with a golden middle to it. And sure enough it is the first Michaelmas daisy. That is the really most autumnal of all flowers, just as the primrose is the most special flower of Spring.

The second of the three happenings belongs to the moorland. Up on the high moors, where there is a broad flat place with a little marshy pond in it, the Elf and the Imp have a few very special friends. There are the curlews, with their speckly brown bodies and long thin beaks and whistling screams, and the grouse who make a noise like an old clock running down in a hurry when they leap suddenly into the air. But these are not the favourites. The birds that the Imp and the Elf love best of all on the moorland have a beautiful crest on the tops of their heads, and they are clothed in white and dark green that looks like black from a little way off. They cry pe-e-e-e-wi, pe-e-e-e-wi, and the Imp cries "peewit" back to them. Some people call them plovers, and some people call them lapwings, because of the way they fly, but we always call them the "peewits." All through the spring and summer they are there, and it is great fun to watch them, for they love to fly into the air and turn somersaults, and throw themselves about as if they were in a circus, just for fun, you know, and because it is jolly to be alive.

But in the Autumn many of the birds do strange things. Some, like the swallows and martins, fly far away over the seas to warmer countries for the winter. Some only come here to spend the winter months, living in cooler countries through the summer. And the peewits, when Autumn comes, collect in tremendous flocks. All the friends and relations of our peewits on the moor seem to come and join them, and then they move away all over the country from place to place, wherever they can get food. When we go up to the moor just at this time, we see not two or three or half-a-dozen peewits, but crowds and crowds of them flying low, and strutting on the ground, with their crests high up over their heads.

The last of the three happenings is the saddest. Do you remember the haymaking and what the hay was carted away for? You remember how the farmers stored it to feed the cows in winter. At the end of the Autumn comes an evening when the cows are driven home for milking, and do not go back again. The fields are left empty all the Winter, while the red and white cows are fastened up in the byre (a byre is a nice name for a cowshed) to eat the hay. When that day comes the Imp and the Elf always walk home from the fields with the cows, and pat them and say good-bye to them at the door of the byre, and promise to come and visit them during the Winter. And then they come home to the house, and knock sadly on the door of my study, and come in and say, "Ogre, the cows have been shut up for the Winter, and nurse says we are to begin our thicker things to-morrow." And then we are all sad, for that means Winter. And I have to tell all sorts of jolly stories of King Frost and the Snow Queen before we are cheerful enough to go to bed.

IV

WINTER

In Winter, real Winter, we get up with our teeth chattering to tell each other how cold it is, and we find the water frozen in the basin, and the soap frozen to the soap-dish, and the sponge frozen hard. That is what Winter is like indoors, and it is not very nice. But there is a nice indoor Winter too, when the fire is burning in the study grate with logs on it from old broken ships, making blue flames that lick about the chimney-hole. The Imp and the Elf plant cushions on the floor, and I sit in a big chair and read stories to them out of a book or tell them out of my head, making them up as I go along. That is the greatest fun, because I do not know any better than the children what is going to happen, whether the green pigmy or the blue will win in the battle in the water lily, or whether the little boy with scarlet shoes will be eaten by the giant, or whether he will make friends with him and be asked to stop to tea. We can make the stories do just what we want, be happy if we are happy, or full of scrapes if we are feeling naughty.

When we are in the middle of stories like this, we hear a tremendous scream-ing, screaming, screaming outside, and a white cloud passes the window with a great, shrill shriek, and we all jump up crying, "The gulls, the gulls!"

And in the meadow and in the garden and flying in the air, screaming and laughing with their weird voices, are hundreds of seagulls, blown inland from the sea, bringing wild weather with them. You know the place where we live is only a few miles from the sea, where it runs up into the land in a broad, sandy bay that ends in wide marshes. There are seagulls in the bay all the year round, and we sometimes see them in the fields in Summer before the storms reach us from the west. But in Winter when the cold and windy weather is coming, they fly in great flocks like clouds of huge snow-flakes, and we watch them from the window and wonder how soon the storm will follow them. And the next day or the day after, or sometimes the very day when they come, the air is white again, this time with driving snow. It comes flying past the windows, to be whirled up high by the gale and dropped again till we see the ground speckled with white, and then white everywhere except close round the big tree trunks. Even the branches of the trees are heaped with snow, so they look like white boughs with black shadows beneath them.

It snows all day and all night, and when our eyes are tired of looking at the shining dazzling white, we come away from the window and sit down by the fire, and talk about it, and think of children long ago, who used to tell each other, when it was snowing, that geese were being plucked in Heaven.

The Imp and the Elf put the matter another way. The Imp says, "It's old King Frost freezing the rain, isn't it, Ogre?" I say "yes." And the Elf goes on, "He does it because he wants to run about and play without hurting the poor little plants. He knows that he is so cold that they would die, like the children in the story book, if he danced about on top of them, without covering them with a blanket So he just freezes the rain into a big cosy white blanket for them and lays It gently down."

Presently, after we have been talking and telling stories for a little, the Imp cries out, "Ogre, Ogre, we have forgotten all about the cocoanut," and the Elf shouts, "Oh yes, the cocoanut," and away they fly, leaving the door open and a horrible draught in the room. But soon they run back again, with a saw and a gimlet, and a round, hard, hairy cocoanut. We bore a few holes with the gimlet, to let the cocoanut milk run out. The Imp likes cocoanut milk, but the Elf hates it, and says it is just like medicine. Then comes the difficult part. I have to hold the cocoanut steady on the edge of a chair, and saw away at it, all round the end, while the Imp and the Elf stand watching, till the hard shell is cut through. Then we knock the end off and the cocoanut is ready. Ready for what, you want to know? Look out of the window and then you will understand. All the ground is covered with snow, and the poor birds are finding it difficult to find their food. The Imp and the Elf, who love all live things, and the birds above all, could tell you a little about that, for every winter day, as soon as breakfast is over, they collect all the scraps off everybody's plates, and the crumbs off the bread-board, and throw a great bowl of food out on the snowy lawn. And then there is a fine clutter and a fuss. Starlings, and jackdaws, and sparrows, and blackbirds, and thrushes, and sometimes rooks, and once, one exciting day, a couple of magpies, all squabble and fight for the food, and of course the sparrows get the best of it, because though they are so small they are the cheekiest little birds that ever are. When all the food is done the birds fly away, and leave the snow covered with the marks of their feet, like very delicate tracery, or like that piece of embroidery that the Elf is trying to do for a Christmas present, when she is not busy with something else.

Well, well, but still you have not told us what you want to do with the cocoanut. Wait just a minute, just half a minute, while I tell you about the robin. Little Mr. Redbreast does not let us see much of himself in summer, when he is off to the hedges and the hazel woods, having as gay a time as a happy little bird knows how to enjoy. But he is a lazy small gentleman, and as soon as the cold weather comes, he flies back to the houses where he has a chance of scraps. He even flies in at the pantry window and chirps at the cook till she gives him some food.

There are some other little birds just like the robin in this and these are the tits. In the Summer we can see them in the woods if we go to look for them, but they do not trouble about repaying our call; they do not come to our gardens very often. But when Winter comes things are on quite a different footing. They are very fond of suet or fat or the white inside of a cocoanut, and as soon as the snow comes so do they, looking for their food. We tie the cocoanut up with string and hang it outside the study window from a big nail, and before it has been there very long there is a fluttering of wings and a little blue-capped bird with a green coat, blue splashes on his wings, and a golden waistcoat, perches on the top of it. He puts his head first on this side and then on that, and then he nimbly hops to the end of the cocoanut, just above the hole, bends over, and peeps in. He flutters off into the air and perches again, this time in the mouth of the hole; and then suddenly he plunges his head in and has a good peck at the juicy white stuff inside. Presently another blue tit comes flying, and then another. They perch on the top of the cocoanut and quarrel and flap about till the first tit has finished, and then they both try to get into the hole together and find that it is not big enough. We all watch them and would like to clap our hands at the performance but dare not for fear of frightening them.

At the beginning of the Winter the tits are very shy, but later on, if the window is open, they often alight on the window-sill and have a good look about the room when they have had their turn at the cocoanut and are waiting till the others have done to have a second peck.

I think all the Seasons are jolly in their own way, and perhaps it is a good thing that they are all so different. Do you remember the Autumn fairy story? Well the Seasons really are just like a family of sisters, and we should find them very dull if they were all exactly the same. After the snowstorms, when we go out together, things are quite different, and we are quite different. The Imp and the Elf wear red woolly caps instead of sunbonnet and straw hat. They wear thick, fluffy coats and piles of things underneath them, and thick furry gloves. Why, the Elf carries a muff just like any grown-up. And the ground has changed as much as they. It is all white with snow, so that it is difficult to believe that the hayfield where we played in Summer is really the same place. We put on our thickest boots, and they go crunch, crunch in the crisp snow. And we gather the snow in our hands and make snowballs and throw them at each other. And then we make a giant snowball, The Imp makes the biggest snowball he can in his hands and then puts it on the ground and rolls it about. Everywhere it rolls the snow clings to it, and it gets bigger and bigger till at last it is nearly as big as the Imp himself and it takes all three of us to roll it. We roll another and put it on the top of the first, and then a smaller one and put it on the top of that, and then we roll snow into long lumps for arms, and there we have our snow-man. We make eyes for him with little blobs of earth, and a nose and a mouth, and in his mouth we always put one of my pipes to make the poor fellow comfortable.

When we are tired of making snowballs and snow-men we go out of the garden and across the road and along the field paths to the wood, tramping through the shining snow. And we drag something behind us: can you guess what it is? Do you remember in the fairy stories about the people who lived near the forests? When the winter came they used to shiver and rub their cold hands and go to the forests for firewood. And as there were wolves in the forest they used to take a sledge so that they could carry the sticks quickly back again before the wolves could catch them. Well, when we go to the woods in winter we pretend that we are going to the forests for firewood and we drag the Imp's big wooden sledge behind us, and keep a bright look-out for wolves, though, of course, there are no wolves in England now. All the same it is very good fun to pretend that there are.

A jolly time we have on the way to the woods. The hedges are all bright with hips and haws, coral colour and scarlet, the fruits of the wild rose and the hawthorn. They glitter like crimson jewels in the white hedges, where the birds are eating them as fast as they can. The sunlight shimmers on the snow of the fields and the snow of the woods, and the broad white shining slopes of the distant hills. And, of course, all the way we watch carefully for the tracks of the wolves. We do not find them, but we find the tracks of birds that have gone hop, hop, hop, leaving each time the print of their feet in the snow, and the little paddy tracks of the rabbits, and the flap tracks made by the rooks' wings as they flag up from the ground.

For a long time the road is all up hill, but then we come to a deep slope down, when the Elf and the Imp sit on the sledge, and I give them a push off, and away they slide, quicker and quicker all the way to the bottom; and then, instead of going straight on to the wood, they drag the sledge up and go down again, and then once more, and then we all go down together and sometimes end in a heap on the snow.

When we leave the sliding hill we go up into the woods, and sometimes really do find the tracks of a big four-footed animal. The Imp and the Elf cry, "Wolf, wolf," but we know that really it is not a wolf, but a big red fox with a bushy tail, who has passed that way in the night, perhaps after stealing a chicken from the yard of one of the farms.

The woods are like fairy woods now, just as if the fairies had hung them with glittering jewels, for they are covered with snow and frost, and icicles, too, when the snow on the boughs has begun to melt and then been frozen again. We hear crunch, crash, crash, crunch, and then the woods are very still for a moment, and then we hear a great heavy crunch, and perhaps see a mass of caked snow tumble off a branch to the ground.

If it is near Chrismas time we do not bother about looking for sticks and dead branches. We walk straight along the edge of the wood to where three stout holly bushes grow close together. You cannot think how pretty they look, with their dark green leaves and red berries, and the white snow resting on the leaves, and you just cannot think how prickled we get in picking the branches of holly. But we think of Christmas fun, and do not mind the prickles much, while the Elf sings:

> "Get the pale mistletoe,
> And the red holly.
> Hang them up,
> Hang them up,
> We will be jolly.

"Kiss under mistletoe,
Laugh under holly,
Hang them up,
Hang them up,
We will be jolly."

The mistletoe is not so easy to find as the holly. I remember once after we had piled the sledge with holly and dragged it home, the Elf pouted her lips and looked very unhappy and said, "There isn't a mistletoe tree anywhere in all the world, not even in the Long Wood. We shan't have any mistletoe for Christmas." Things really did look rather sad, so I sent her off to ask the old gardener about it, and the Imp went too. In about an hour they came running back all smiles and happiness, and with their arms as full as they could be. I shouted out to them as they went past the window, "Where did you get all that mistletoe?" And they laughed and said that it grew on the apple tree in the orchard, and that the old gardener had cut it for them, and promised to let them bob for apples in a bucket in the wood-shed if they were quick back. That is really and truly the way the mistletoe grows. It is just like a baby that will not leave its nurse. It pines away and will do nothing by itself, so it has to be stuck into another tree to grow there more happily.

But you know the snow does not last all the Winter. After Christmas, and before, there are weeks without any snow at all, and then we find it rather sorrowful to walk over the bleak bare fields. But the hips and haws are bright in the hedges, and whenever there is sunshine everything is made jolly. And then, too, it is great fun to watch them ploughing the land ready for next year's corn.

"Old Susan isn't pulling hay carts now," says the Elf, and we look up and there is Susan side by side with another of the farm horses harnessed to a plough. And a boy, a big strong boy, holds the handles of the plough and the reins at the same time, and shouts to the horses, and they cross the field slowly, tramp, tramp, tramp, and the rough earth is turned over by the steel ploughshare, all dark and earthy, ready for the seed. In the middle of one of the fields is a special friend of the Imp's. He wears a battered hat and an old green coat, and a red worsted scarf that flaps in the wind. He is made of two sticks, one stuck in the earth and one nailed across it, and he is called Sir John Scarecrow, because it is his duty to scare the birds from the field. But we have laughed many a day to see the rooks perched on his broken hat and tattered arms.

When we think of sowing seeds we think of Spring with the new corn green on the red ground, and when we think of Spring we think of Summer, when it is tall and wavy in the wind, and when we think of Summer we think of Autumn when the corn is golden and cut, and then, why, then we come to Winter again. And now the Imp and the Elf say that I have told you enough about our Seasons, and that I must tell them fairy stories till it is time for them to go to bed. So here is good-bye to you and a piece of advice. If you have got any grown-ups near at hand and it is not quite bedtime, if I were you I would ask them to tell you stories, too.

Lector House believes that a society develops through a two-fold approach of continuous learning and adaptation, which is derived from the study of classic literary works spread across the historic timeline of literature records. Therefore, we aim at reviving, repairing and redeveloping all those inaccessible or damaged but historically as well as culturally important literature across subjects so that the future generations may have an opportunity to study and learn from past works to embark upon a journey of creating a better future.

This book is a result of an effort made by Lector House towards making a contribution to the preservation and repair of original ancient works which might hold historical significance to the approach of continuous learning across subjects.

HAPPY READING & LEARNING!

LECTOR HOUSE LLP
E-MAIL: lectorpublishing@gmail.com